THE RATTLING WINDOW

The
Rattling
Window

Poems

Catherine Staples

![logo] THE ASHLAND POETRY PRESS

Printed in the United States of America

ISBN: 978-0-912592-96-1

Library of Congress Control Number: 2012949209

Cover art: *Collection / Olympic Flame Tulips,* David Graeme Baker

Front cover design: Ed and Alina Wheeler

Back cover design: Nicholas Fedorchak

Interior layout and design: Sarah M. Wells

Author photo: Ed Wheeler

Acknowledgments

Acknowledgment is made to the editors of the following journals in which these poems, some in slightly different versions, first appeared:

Blackbird: "All Souls Crossing"
Centennial Review: "Night Feeding"
Commonweal: "My Neighbors' Pools," "Anchored and Listing," ("In the Woods Hole Harbor"), "With Trawlers and Fishermen" ("On Black Pond"), "Legolas" ("Ten Year Old"), and "Dwelling"
The Cortland Review: "Where the Hill Rises"
Michigan Quarterly Review: "Impulse" and "Steeplechase"
Nimrod: "Wild White Curtains" ("Summer Cottage")
NOW! (then) Anthology: "Watchers" ("At the Oysterman's Cottage")
Poetry Northwest: "Salt" ("Changeling")
Prairie Schooner: "Wyeth's Anna Kuerner," "Seafarer," and "Fear of Heights"
Quarterly West: "Napping on the Good Rug" ("Lurchers Napping")
The Recorder: "Flying Change" ("Clark's Field") and "Mother, Daughter Before the Madonna dell'Orto"
The Southern Review: "After the Storm," "At Tea," and "Into the Blue" ("Smith's Cove")
Third Coast: "What I Remember"
Valparaiso Poetry Review: "Wintering Over" ("Hacking Out")
William and Mary Review: "Caretakers" ("Boot-top Farm")

Under the title "Intimations," "Edgeless and Smoothed Out" won an honorable mention in the Morton Marr Poetry Contest of *The Southwest Review.*

My gratitude to Villanova University for support in completing this manuscript and to Eamon Grennan, Peter Fallon, Dean Rader, David Moolten, Margaret Holley, Nathalie Anderson, A. V. Christie, Joan Houlihan, Fred Marchant, M.B. McLatchey, Catherine Prescott, Evan Radcliffe, Nancy Shaw, and Natalie Staples. I thank them for their close reading and sound advice. I am especially grateful to Stephen Haven and Sarah Wells at The Ashland Poetry Press. My sincere thanks to Ed and Alina Wheeler and to David Graeme Baker for permission to use his beautiful painting, *Collection / Olympic Flame Tulips.*

for James,

Claire, Natalie, and Kevin

Contents

Coming & Going

Into The Green

Paint Like Breath on a Pane of Glass

There are no worlds but other worlds
And all the worlds are here…

— Wendell Berry

I prefer winter and fall, when you feel the bone
structure in the landscape—

— Andrew Wyeth

COMING & GOING

Fear of Heights

A widow's walk will go to your head—like the sight of a former
boyfriend pulling up in a two-toned Alfa, sunglasses
and a baseball cap, he patiently waits while you study his face.
Recent history you can't know but might intuit—beaten up
some by previous inhabitants. Still you remember.
Twenty years or more, the patina's all glint and shiver.
It's the wind from the sea makes you light-headed,
inclined to break like a floe far to the north,
present self sheared loose from your youth.
What you have in mind is nothing,
walking around the porch to the back door,
the half-filled lemonade pitcher spilling phlox.
The latch unhitches to the drop of a thumb
and summer rushes out with a long-held breath.
How easily sheets fly off the wicker—chairs,
tables, dining set, a summer writing desk.
Like the arrival of guests from all doors at once,
the empty room is busy again breathing in the sea.
You try one door, then another, quick as the mother of pearl
spill of buttons, you're there in the stairwell looking up
into impossibly bright lines of light edging the hatch.
With a shoulder and palm it'll heave loose—
already the salt air, the unguessed dimensions swaying
in wind. The real weight of the door and the sheer
white of the height, like sun flashing dizzily over the waves,
the bright likes of which once caused a boy to fall to the sea.

Flying Change

Bright as it could be past March—
the boggy thatch that might sink
a pastern gone firm as a spread palm
and blindingly green—you rarely

lost track of the dark in Clark's field,
not with all that border oak and fir.
Somewhere up there our postman's
father did himself in.

In which of the four fields we never
knew though the morning shot—
we must have heard it while heaving
dung into a barrow, shaking out

clean straw, breaking the iced tops,
steaming the mash. We must have taken
it absently for hunters out of season,
a baulking combine stalled, then quit.

Schooling horses in the long
grassed tracks, you wondered could this
be the place, the center around which
our walks unwound into slow trots, then

barely held gallops, that wild straining
at the circle's taut rim that in hours
and months settled to rhythmically, cantered
halos. Perhaps it was there—at the still point

of figure-eights, where one loop ceased, another
began. If there, spirit forsook flesh
for thin air, and if in a lightheaded instant
the horse unknowing, lunged onto the changed lead—

unsettling leaves, loosening rooks
from the listing, dark-eyed firs
with all the sharp violence of an unforeseen,
and irredeemable, second thought.

Anna Kuerner

After Andrew Wyeth's "Groundhog Day"

The painter who wanders your house night and day,
sketches his way in and out your back door,
kitchen, barn, and milking room, he's erased all trace
of you. Look, your favorite tea cup, the one
that's snug to the curl of your forefinger, even that's gone
bleached white as moths, something a dream tossed back.
The print has disappeared from the china rim—no more
apple bloom or trumpet vine vining a smooth weave.
No gray green, salt blue, faint as any wave
glimpsed from distance. Day after day
he paints you and the dog sleeping—shut eye, wolfish
set to his jaw—then the bunchbacked skittery quick.
But in the end, the dog disappears as you do.
Cup, knife, plate. His fangs menace from the rough-cut
log beyond the open window. And you? Are you the strip
of light glancing the wall, obstinate refusal to quit
or give in? It was your hands set the table, raked the grate,
chopped firewood far side of the pines. Is it your quiet
the painter caught? The long slow place before the scrape
of a gate lifts from its hinges and your husband strides in
fresh from New Holland, rushed talk of horses, calves, and tractor
gusts the room like an unseen wind, settles to the porcelain
chink and domestic sing of knife on a dinner plate.

Caretakers

It's the place we still dream of—under the hilltop.
White stone, clapboard—house full of windows.
Wind and light crossing like geese, one room into
another. Bolts of violet iris in the midst of the stream.
And late at night walking, watercress small-leafed
and fine as moss, islands of it, rooted yet moving
on the water's dark skin. Snakes, too, in dank sheathes
under the stone bridge, deer paths leading nowhere,
and the unlit place marked the anger of our first fight.
Gloom that might have set the woods on fire with words unsaid
but didn't, and then there were deer lifting their heads.
Forgiveness and wind-cracked sheets. Nothing to keep us
from sleeping 'til noon, until the pasture hill was white light.
And the pond, again, thrummed with bullfrogs, skimmed gauze
of waterbugs, dragonflies. The kettle whistling on and on.
And after we'd gone, honeybees stormed the open window,
hummed themselves to stillness near the furrows of our sheets.

Anchored and Listing

You can't see them from Shuckers, even
the Captain Kidd, not for the sea of masts
rocking at anchor, Lasers and Snipes
angling in on changing wind. You have to cross
the drawbridge where the whole road lifts
for the passing of a single sail, wander
downhill past missing pickets and gray
disrepair to the harbor's backdoor. There
in the bilge and muck, eight to ten
spent rowboats and a tipped bluewhale of a dory—
bleached, peeling, someone else's summers
disappearing in weeds. No more remarkable
than a neighbor's sheets idling in wind.

But once, just there, in the oily trap by the dory's
stern, two boys vied for turns at the tiller.
Their sister trailed a branch in the rilled wake,
felt the light thrum of resistance.
Father talked reaches, runs, heading high
and low—then with one deft tug, reeled her in
close to the rocking red buoy. Their mother looked
leeward—blue kerchief bellying wind
like a tight jib—she daydreamed and napped
behind dark glasses. And the sister looked up
to see an island becoming itself—sandy spit,
and a slender white lighthouse with a catwalk above
and someone there—a man?—no, a boy
on the jetty, centerboard humming, a lone skate
rising out from the sand and little more
than an arm's length away—a dark-eyed boy—

sleek as a seal with seaweed and water, hair spun loose
in an arc of gleaming flicks, the unruined wild
of his eyes—still as a hare in the lamp of her gaze.

My Neighbors' Pools

The way seals ride an incline and sweep
like light down scoured tanks,
you know they're dreaming the sea.
Nights like these—windless, landlocked—
all I need is a wintry descent in someone's pool:
crescent or quadrangle, clotted
with leaves or made fast with tarp
that icy slip into otherness,
chest, then hips grazing the pale slopes
of the pool floor. Numbness like a second
skin, eyes gone dark as a seal's
trespassing in the dead of night, out of season
in my neighbors' pools. Never the same one
twice in a fortnight, I've grown meticulous about latches
and gates, the loosening of knots, the first
lifted corner where the moon rides and the waters stir.

Arms folded close as fins, I rush the blackness
headfirst, hunt the circumference of each new pool,
dark overlap of leaves in unlit corners, chubby
rounds of crabapples, spiny feel of crickets.
Through the chill waking, I grow sleek and fat
in the frigid blue hold—so buoyant it seems
my flutter kick is all spine. I plummet and rise
through the underside of autumn, scarlet bleed
of maple on oak, the skin-taut border of water and air
christened in cold and shining with dark leaves.

What I Remember

It's not the woods so much as the path, the rutted
track past the Brown's. Their chicken coop hitched
at the corners with baling twine. Listing, rusted-through,
and still it kept them in—the bantams.
Small eyed as dark gems, restlessly mean.
They'd just as soon stab your palm as accept
whatever it was you slipped past the mesh. And Mr. Brown,
the dull hammerhead of his quiet, moving among them.
The swung crack of firewood piling beneath him.

You ate on a stool in that house—odd things, silvery
fish from peeled back tins. You kept as quiet
behind your bangs as Mary Brown herself.
Past kindergarten the divide between us was clear—still
there was the path by the house, only way to the woods.
You tried not to stare, but the otherness of them
drew you in. The mother's turned back, the one brother
lost as a daw in his own yard, the others bare-chested
under hoods and ripping away at engines and mowers.
The older girls red-lipped and world-wise, already
going somewhere. You tried to keep your eyes reined in,
resisting so much as half a stare. But all the way to the fields
they held you like the one full mirror in the run of the house.

Previous Owner

Not much good with any of his wives.
Rock to their stream in all weathers.
One sat here in the wingchair,
Shrill needles clacking a racket.

He kept a goat by the spring house,
Tethered to a rusted stake; that goat ate
Up the bright eyed primulas, then
The hooded Jack-in-the-pulpit, all

Grass to speak of. Eyesore, grazing
Ruin. The very sight of the white goat
On the steep slope bore a hole
In her spinstery heart.

Which is why he loved that goat,
Banked hay for him in the barn.
Up the rickety steps for sweet
Sheaves of timothy, summer distilled

In armfuls borne downhill. He missed
His old field hunter, busted up
But lifting his head, shaking his mane,
That cheery trot to the gate.

He ought to have retired him
To their long meadow.
It might have been apples
And clover for the old campaigner—

Not the slow mill of the Snow White
Laundry on Goshen. No siree,
He'd give no ground on the goat.
He was wised up

But good to her wheedling. Which
Is why he didn't so much as flinch
When she stepped out with the Winchester.
He'd seen her take shots at groundhogs before,

Those wooly rugs heeling it double –
Quick, tumbling like waves into the wood.
She wasn't much good, but she nailed him:
Plain daylight, middle of the drive.

Shot him twice for good measure.
Shoulder, then hip, the old coot.
You can still see the chipped bit of brick
By the spring house window.

But spite's got grit and the old man survived,
Lives on yet in the face-off of twin fireplaces
In the front room of the old farm on Providence.
In the invisible quiet, she stares daggers

And he dozes oblivious, serene in the faint hum
Of honeybees searching out a new queen,
He sleeps and dreams the bearded white goat
Loose on the front lawn, ceaselessly chewing.

Steeplechase

No lush beauty, but a sloe gin
dark bay like pine bark in rain,
all angle and ridge. One long look—
and I break like a thief through the maze

of their legs, past tweed skirts and trousers
into the musk closeness of strangers,
slipping sideways through the gap
in the storm fence, slipped away

and close now to the smooth gloss,
oh, quick knots like fast water
go shoulder and hock.
Glittering black, bay, chestnut

coming up at a jog, crush
of velvet on wool, nearly there—
horse with my uncle, *Waitawhile*, wait,
just as someone's bony thumbs

go under my arms, all horses
and riders tipped at an angle,
backwards up in the air, all
of them moving off, hooves

ringing in pairs, in triples, thunder
from the earth. I am hoisted
up onto someone's broad shoulder
as the racehorse rises into the blue;

divots fly arcs over the beaten ground,
then the sound of things returning, the field
grown small—bright silks & caps
and flaxen tails through the gap and gone.

At Tea, Far Side of the Glass

To be grown up is to sit at the table with people who have died,
who neither listen nor speak —Edna St. Vincent Millay

I sit at their table, and my dead speak to me
Sometimes. Handsome boy with dark brows—
He died while I was in England, letters
Crossing in onion-thin blue skins,
Mine, his, mine, then not his—
We parted in a dream and now sit
At far ends of the table and stare, mouthing
Words to songs, the same songs. Once
At dusk he stretched out lonely
As an oak, shadow just nipping the ledge
Of the bedroom window.
The dead are as curious as we are—
They haven't all the answers.
He wanted to see if the child looked like him.
Imagine that wistfulness—when we'd never kissed.
He rose to the glass in a breath of frost,
Watched the boy sleep, snail shells of his fingers
Lined against his cheek. Susurrus
Of waves in each exchange, skittery rush
Of side-walking crabs into grass.
I've missed him so—
And wondered why he didn't stay—couldn't,
Wouldn't, or not allowed—but the kingdom
Of the dead is a lockbox of mystery.
And these are not the things to broach at tea.

When my grandmother died I listened for the wind
To come howling down through the chimney.

At last I heard her but faintly. *Each child is another island,*
Another rough sea calming, west wind dying.
And the coracle turns, turns, and then glides smooth.
Without so much as an oar, home again, Galway.
She called and called, but I would not leave—not
With the children breathing lightly under their printed sheets.
There were days I thought I'd slip off of an evening
And return after the owls and just before the doves,
But now in time we settle into the routine of the dream.
She loves this late blue hour, raises her tea cup up
To see birds in ivy twining the rim—bone china,
And the tea is amber, fresh with a hint of rained-on hay.
My grandmother sits at the table, lovely braids
Gone blond again and her fingers light on the white cloth.
She recites the verses she learnt as a girl, the part
About the barge and the mirror and the weaving, & the water
Of course, her memory of the flooded bay, *why the sea takes us all.*

Now it's time, my grandfather politely looks away,
Spoon laid neatly, ringing an ending on the china plate.
He slides out his chair with a violin's scraping,
And leads me outside with a great gust of light
To the espaliers, where I look back through glass
To see the others still at the table. We cross
To my own backyard through apple, pear, apple.
His hands guide mine. He thinks something hard
And I listen to the sparrow and the grass, then I thread
The baling twine in a skater's loop 'round the branch
And the air and the guide wire, draw it in taut
But slow. And the branch bends; the idea in his mind

Takes shape in my tree. His voice is clearest, *Kate, Katie*
He goes on all morning with the chickadee
Hopping in clematis—that dead-as-a-doornail tangle
Of brown that'll bloom soon, rioting violet
In glorious vining tents of green.

Napping on the Good Rug

Back from new cold of January, two hounds curl up
with the silver fox, hiss of skates into corners
on the frozen pond. Brazen pause of their quarry
and they're off again in their sleep,
squealing cross the pasture.
And it might be spring in the dream,
water running above the sap house,
brown hare waiting by a gap in the wall,
head tucked to the ruff of her flecked coat.
and just as someone crisscrosses uphill—
her bright eye stirs, the darting mercury of boys
start her up into the steep scree and shaded islands
of ice. They wake on the patterned rug
having been somewhere exotic yet familiar,
the sharp hot ride circling in,
hair-pinning back like the procession of a local saint
carried aloft through narrow streets
impossibly slow, faintest scent under incense
on cobbled stone—then in a clap
they're off into the Venice of their dreams:
a remote campo in San Giobbe teeming with cats.

Legolas

What I love is that loose as a tossed wave
posting you do, sweet chestnut making fast work
of the Sugartown flank of the hundred-acre.
You thump his sides for a fast trot
and he swings loose for you in long arcs,
hooves sweeping lines through the last
of the snow. We're moving along now,
you astride and me on foot. You are Legolas,
Tolkein's elf, and I your devoted slower counterpart.
Light as air with your fixed dream, already you spot
the path through pines to Black Sheep on Garrett.
You've no worries about jeeps and pickups
roaring the long straight break Sugartown to Goshen,
and I've no more sense than your sister
as we cross in the dark—until I see an old friend
at his mailbox. He leaps the tumbled wall
to make your acquaintance, leads us past
snowy garden & shed, huge gray drafts,
breaths steaming clouds, stamping
and peering from their stalls. By the time
you hear the host of red-brown hens humming
and clucking like the sing of water on stone,
our wandering's turned journey. We might
be alone amid crags in blue mountain mists
with this last gift: two brown eggs, speckled
night maps of another world's sky, the living
weight in my cupped palms all the long jog
home through the rhythmic, snow-lit dark.

August Storm

Big weather bellies up
north of the old schoolhouse
and Rock Hill, sheep gather
in the far corners, tucked in
over knees and hocks.
They make themselves small.

Sharp cracks split the sky.
This morning's lazy
crosshatch of cicada
ratcheting up and down
is another country
far from here and now.

The dark comes on
quick as crosswind,
the full measure
of a week's rain emptying
in minutes, our shirts and shorts
plastered to us, catching—

Our rush is slow,
we're fumbling
with the splintery wood
of slipboards not yet worn
in pasture gates, latches ring
and lift. The horses strut.

Tails up, necks arched,
a terrible rippling

shocked white
in lightning's strike,
hardening like
Cararra marble on the hillside.

Some dire awakening;
then the light goes
they're at the gate,
forelocks and manes dripping,
snorting in the charged air,
glittering and chastened

as lightning moves off
towards West Chester.
A hundred green leaves
blow loose of summer—
and the pasture
takes back the dark.

Atropos & the Goldfinches

Look at the one with the shears,
head down, intent
on the thread she attends:
life spun, measured,
cut off. She must close
in the precise place—
that grass thin hit of light—
now, not a moment later.

Nothing vindictive
in her pure face,
she is young and earnest.
Look with what grace
she leans, foreleg
taking the weight
while a lawn mower hums
and a half-dozen goldfinches
sway on stems of coneflower,
patiently, each seed,
pivot and tip.

A brother and sister
kick a ball. She's close by,
but they can't see her;
she's the dizzy blind of light
at the garden's end,
fine net of midges
at the compost's rim.
She is who she is,
and it's all one to her:

mayflies in a day,
monarchs a few weeks.
In another breath she'll step,
iron blades will meet.
Some living breath will cease
and a loss will fall home.

This afternoon—
it's the children's pony;
tomorrow, your father or aunt,
someone's dearest dear
will slump from this world.
The glittering thread will fall
in a tangle of autumn knotgrass,
fluttering its last magic
in the nave of the children's palms,
insufficient solace
for the next night's grief.

Edgeless and Smoothed Out

Maybe you're gracing another place—
milkwhite bracts of dogwood blazing

amid privet and honeysuckle. Night air.
The dogs taken by whim on a downhill tear.

The bay's in his pasture wearing his fly mask
like Zorro but nobody notices, nobody grins and asks

Did you see him just now? And look at the cat
trailing dross—in cobwebbed glory—that fat

orange cat lords it over the dogs by the tack room
tap. Sun's high and horses will be in soon.

The freight train stops in the middle of the woods
and we don't know why—

Then it starts up again, and the orange tabby, regal
as Bottom in his donkey dress steps down in full

possession of the three dogs' attention, and they all bark
but none dares disturb his progress, the stark

procession of morning where everything goes on
as before, but nothing's the same.

Earth and Sky

After a blizzard like this—how bright
blinding white are open fields and pastures,
like walking into a future untranslated—
just the faintest rill on surface under which
a tractor rut rides. Draped lines of a covert
where once an owl blew by on smock-white wings
so close and silent we watched him rise
in a slow rowing motion, air thick
as the stupor that held him, dreaming still,
but functioning in the instant. Blown sweeps
deepen in hollows, far as the eye can see—
an unbroken shimmer. We squint hard
into dizzying white, the swing of small hills,
but these ponies know the way back surely as the spill
of carrot and sweet feed drumming a bucket.
They improvise a leaping, high stride to carry
us home, vaulting each dare-me-die glitter
of snow melt and bare earth as if it were a hidden river,
the glimmering dark Styx itself and her greedy
boatman all too ready to steal us from this beauty—
gap in the path, fissure in the squall,
earth and sky mixing,
and an open gate swinging noiselessly, swinging
on its hinges.

INTO THE GREEN

Thinking About the Sistine

Rise. It's hard to imagine on a twenty degree day,
the blessed rising from their boats
all along the rugged shore, even the rocks
borne heavenward in Michelangelo's vision.
Such unlabored rising is far beyond our ken—

Take this red-tail feasting on our neighbor's rooster,
one bright feather at a time, pink flesh at the nape
of the neck stringing loose, matter-of-fact shake
and tug—there's something we know.
The dark drop, loss, lithe brightness strewn
the length of meadow and the children next door
still calling out under henhouse, shed, and tractor.

Although I change my skin, Michelangelo wrote,
in a sketchbook under the scaffolding
as if *the last day* were a simple undressing,
seasonal, no more astonishing
than the whip-long, honeycomb casing of a snake
cast off between logs in the woodpile.

The Breathing Green: An Elegy

I. The Farmer Shocking the Corn

That morning sun flashed
abundance and the heady bloom
of something was loose
among the river trees.
Wind caught and blew—
and the train came on
from far off. Everywhere—
the warmth of cut corn
as we scythed and laid the stalks.
My long curved blade
a wonder still—
"Keep the wood-piece in line
with your shoulder.
Two hands on, sweep
a long stroke—the blade
will slice hay like light."
My father's words came back
as the painter and his grandson
stepped from the station wagon,
walked toward us.
"A lost art," he told the boy,
"You won't see it much longer."
We showed him how to lay
the corn in shocks:
twelve together,
six to a side, held upright
by a loose braid of leaves.

The boy stared,
turned towards me—
alertly intent.
It wasn't long after that.
The two of us fairly flew,
and the light was gone from the day.

2. The Painter

Daydreaming, I suppose,
as my grandson and I drove
the Ring Road to pick up Evelyn.
Mind running on—the hillside
laid bare under downed stalks;
summer distilled in burnt
kernels, freckled husks.
A whisper and a thirst—
a long-gone day returning.
We'd walked hours, took turns
at the pump's handle, pressed in,
jerked up—again
and again 'til a breath-slow
wheeze of air answered—
a stone cold deep draught
thrumming upward.

Four of us apprenticing
with Pyle, we cut out early,
paint spattered,

rammy as colts
loosed in a wide pasture.
A helmeted red pump—
middle of the field.
We dropped to our knees
pulled off shirts, tilted head back
to gulp the free run—
underground rust, sweat
and earth, a salt tang—clear
and cold as stars.

I never heard the shunt and thrum.
Imagination drives sound
to silence, habit of the work,
long days of painting.
Gasp of air, sluice of water.
Quiet the world, the heart, the mind
'til what's there, isn't there:
A mass of silver bearing down,
the car going nowhere
gas pedal slammed still—
Iron bars of the cowcatcher.

Then a great din
broke open the wood
one breath, another
quick as birds
into the long quiet
above brown hills
weightless thin

blurred as through ice
but don't look
darling don't
worse than any pheasant
thrown from the tracks
wet moss, dry leaves
blowing in at the door.
In broad daylight
a current of light
humming like a brook.

3. The Meadow

Perhaps the unscheduled
mail train making her rounds
couldn't shake the fields
from his head? Who knows
what's heard or unheard:
wind in the oak,
dogwood, apple
whispered rush or pause,
a long swell simmering uphill
under finches insistent.

It means nothing
to the black racer tracing
fat ribbons through standing corn,
even less to the boy
at play in moss & acorns.

One gust and his world shivers
into the green wood unseeable:
acorn caps, stick-and-leaf pastures,
a spider web winks,
in a dazzle of light.

Daydream's clear as the run
of stones in shallow water,
flecked gray, small and red,
red, ivory and green.
But fate is a confusion of winds,
hillside of scree, remnants
washing up with each rain,
a rusty rim, shards of glass and broken
nails, then a cobalt blue vase,
the color of stained glass in morning,
perfectly intact.

And when the gate
closed noisily, accidentally
on grandfather and grandson,
it opened for the son to paint
what he loved and what was gone.
And the train came on.
Something was wrong
with engine or piston,
someone was mowing. Birds sang
their ordinary refrains. An arm went up

protectively before the child's face,
glint of glasses,
and the scythe did not pause in its whispering.

4. The Wind

Let the pasture be—
ragweed and horseweed
will lay siege.
Later, aster and goldenrod
will bring forth
red admiral and viceroy,
the aspen with its quaking leaves.
A red fox will move in
where a groundhog's been—
one entrance on the meadow,
backdoor to the creek.
One small thing
into another.

Birdsong shakes loose
for the warbling creek,
hour by hour
changing the songs.
A farmer and his wife
stop to listen, a boy
wanders close, watching
clouds ride the water.
Here a scout once crouched

and a musket left its print,
where the voices went
and stories ended;
a hundred undreamt things
in the rill of the running
water, in the singing trees.

N.C. and Newell Wyeth, October 19, 1945

The Fifth Field

A red fox might go far in those uncut woods—
neat prints tracking the field in perfect silence,
shiver of seed heads as he cuts
the diagonal unseen while hounds falter
on the clay path, high-pitched certainty
gone seedling thin while the fox
trots the distance of four fields in two.

Ahead of me on the thoroughbred, my aunt,
her long hair in a French twist
tucked like bird's nest under her huntcap.
Come on now, come up, my pony cantering
to catch her long-strided mare, *Macooshla,*
Macooshla, grain in a bucket shaken like this
will sound like rain. My aunt, not by blood,
but marriage, and for the length of each summer
my other younger mother. *Slip the halter*
with one hand, slide reins with the other,
to call broodmares—halloo
and whooy and knock at the tin gate.
Fill each paired bucket halfway,
elbow in, stride takes the weight.
Ten years later a cold spring,
we stood in the drive
while she explained:
another house and job, pasture where ponies
might graze with donkeys, loose hens;
not to be married any longer—
while we stood on macadam as if
any minute now she'd go.

In my dream there's hardly a farmhouse
from Clark's Field to the stone barn,
then yellow dogs
straining at their leashes tied to a gate,
wild to break free,
hack home with our horses and hounds.
And I look up, forgetting these fields were never
in my lifetime clear of houses,
I'm in someone else's memory of the fifth field,
the full wood and my young aunt. Hours unspent.
The long farewell—just a hunting horn's close
on a good day, innocent noise,
sweet as the shake of grain in a bucket,
broodmares and ponies rushing to the gate.

Soundless Arrivals

Like something out of *Black Beauty* before the squire's son
tumbles low, and the horse is sold. Twenty loose box stalls,
slate roof, and lanterns. Doors to the hayloft
like the great closed eyes of some sleeping god
or python. Held breath, a lid on the quiet life
that must have ceased a century ago. Half doors still
open on lush pasture. A John Deere hums along
the perimeter, and when it wheels south for the long
length, we seize the forelock of Lady Fortuna
and steal in, quick trot, through a gap in pasture along
newly printed tracks and the bright chops of clipped grass
into the old barnyard. We are at the center
of a wood, omphalos of the last vast tract of green
due north of Crum Creek, the slow bend where trout
leap flashing and underground channels muscle
and rush. We take in all the eye can see:
there's a giant hook for levering hay bales,
a brass cauldron for steaming the mash, one stunned arrow
on the gas pump's clock—as if words might save them
from the physics of ruin. It's first look and last
the tractor's thrum doubling—as we slip out unseen
on the chestnut and the bay, in the bright wheel
of the season, apex of sun's arc descending,
clip of their four-beated walk quickening for the wood.

Collapsing Star

The covert edging the Moran's pasture:
ashen green and burnt oak, a hopeless
tangle of vines. The burn pile smokes
and sinks in, a collapsing star of farm clutter.

Charred logs and broken rails, creel
of a muck bucket and old hockey sticks.
Any second now, broodmares might wheel
and buck, trying hard to remember

what it was like before this.
Out here, the mind rides surfaces;
edgy, you weigh the sky, a twist
of leaves. These horses jazzed

by the least thing. Then a dog fox
cuts the circumference of the hundred-acre
like a razory swift, pasture to mud-rut.
And a glint stirs the pond top;

Orion won't forever bend on one knee.
Soon enough above us—the swan,
and milky river, mallards descending
two by two. Stir of dry reeds,

the male's shimmery head,
gliding to view. There
where the water looks haggard,
and the brush bare—

soon enough, it'll happen again:
first the meadow lark,
and then the wren,
sap and curled leaf—

the fleet and breathing green.

The Letter

Walk the path toward the tin whistle of water.
        Sassafras and bayberry, watch your calves
                for the brush of ivy. Deer path perhaps

that's all it seems, until the suddenly green
        banks of moss appear, thick as thieves
                to the lip of running water, the plank across

snapped at both ends—you can see
        where it slipped the bank and fell
            all winter, most of spring.

Now the only way over is in.
        Forswear all fear of swamps, the roadside
                memorial to the Herring Creek shortcut.

Just remember the trick
        with sinking is not to twist.
            Deep breath and another step

grab any sturdy sapling and clamber
        the steep. The path grows less clear—
            look for what seems like last year

summers before, then further.
        Think first transistor, shocking fractal,
            erasing your dark hemisphere.

Follow the deer through a scatter of pines.
　　　　See the two, three, hand-shingled roofs?
　　　　　　Widow's walk gray, upright and narrow

as a single cot, each sleeping porch
　　　　its own island of sunspot and shade,
　　　　　　blue paisley bedspread and feather pillow,

Rilke earmarked with a blade of grass,
　　　　and an unclasped
　　　　　　wren-singing hook & eye latch.

Wintering Over

Just yesterday—four chipped bits of china,
brighter blue than willow; in great arcs
they sifted upward and under their wings
an astonishing umber printed the air.
It was wheel and feint while we trotted along
in long grass, in the loose swing of work
and pleasure on a twenty-degree day,
horses steaming breaths, fingers loosening from knots—
when another unfixed bit of blue shot forth.
I knew then you were stealing a look
at the long waited for, the one
you called *thrush-in-the-hawthorn.*
Surely you paused in the smooth light
of the wild pear to watch—
chestnut hair, sapling thin
her long fingers closing
on the pony's mutinous thought,
give and take, steady leg, half-halt
to a softening hand. All fear sequestered,
disciplined with the alert work of hacking out
in all weathers, her first season following hounds,
heads up in the flying wood, the heart high
blur and burn of tree trunks
and leapt logs into the fierce joy
of the lonely wood and the heart beating hard.

PAINT LIKE BREATH
ON A PANE
OF GLASS

Wild, White Curtains

Perhaps the wind coming up
is just wind, coursing its way
through beach towels and sheets.
But yesterday and last night—
it seemed a mother's blind touch,
she thinks her son
is still here—
but he's gone on
with petals in crosswind,
pair of Friesians drawing the cart,
the long slow walk
to the frozen green at the center of town.

Just us
and the moon,
side by side in the small bed
under the wild white curtains.
They slap flat, then billow loose
undecided as new souls—
whether to cross now
or follow the Herring River back
through the sluice.
But in the instant it takes
for an Atlantic gust
to slam our door flush,
then suck it back, rattling hinges—
something's been here,
and gone.

Feverish whir
in the crackled pocket
of a yellow slicker,
tangle of fishing line
back of the door,
whispery rush of
wind on skin,
spent rose, salt—
the slow flowering
bone to bone, and the curtains
passing over, lapping our heads.

Mother, Daughter Before the Madonna dell'Orto

The empty cups of prayer candles are ruby red, a thousand lira
to fill them. It took several matches, the sulfur burning faint clouds
into the high niches, would we lose our prayers in the time it took

or just this unlikely communion? The two of us kneeling before her,
the quiet of it—knees on stone, her whiteness suffused
and the breathing breadth of those folds in her robe like birth quilts

of a first child, folding and spilling, the knotted primrose, pearled eyes
of anemone, the way you slept as an infant in our bed. Snowfields
to a stonewalls' embrace—did her arm move in the broken place

midway between elbow and wrist, that swathe of plaster less bright
but whiter? Miracle enough that so much of her lost image
should be found and pieced together from the garden-grave

where it was unearthed. No less amazing than this wordless place,
can you feel the clearing—top of the hill, brightest green I am dreaming
for you? My sweetest sweet and first, brush the dark from the folds

of your inside-out shirts, it was only six or seven seeds of the blood-red fruit.
And even Persephone comes back to her mother each and every spring.
Besides which, it's lighting now, the candle glass is blushing crimson,

and it's noon in the jewelbox, a short gondola ride from here.
Already your prayers and mine are emptying from this world
to the next, a moved stone, garments whisked on the invisible wind—

what need have we of witnesses? Come on now, outside the church
your father, sister, and brother are putting away shots
in the chalked-in goals on the piazza walls. Show them what you can do

with your left foot, your right, eyes closed, arms outstretched—
giddily yourself in sunlight. Lean back into the wind, put one away
high and right, just inside the corner of the makeshift net.

Impulse

Foghorn. Open the window, you'll hear it
        beneath the refrigerator's hum and the rain's
light drill. Barely there—the lure of sirens through wax—
        submerged in darkness and rain, miles
on miles of scrub pine, the sleepwalk of rosa rugosa.

Hold your breath—there—the wave slipping wave
        spacing, a summons no longer heeded, not with radar
and running lights. Effortless, serene—tankers keep
        within pulses of red light, white light chambering
the old canal. And any ten-year-old in a catboat knows enough

of currents and wrecks to keep clear of Wings Neck
        by dark. But I've only days, not summers,
and the moon's at the sill. Sleep undone
        by the thunder low sculling. Tide, high; rift
whitening quick as thought. If not for the child—

innocent of whim—tightly curled as a fiddlehead fern—
        but I'd walk slowly. The path to the beach
moss and sand, moon at my shoulder passing
        through firs; no steps would jar, she could sleep.
No current to speak of—only the light shake of rigging

in an offshore breeze. Just as much stir as the float bobbing
        at our dinghy's prow. And this child so much further
than the last, nearly the seventh month. Surely the womb's
        quiet, not unlike a tamed cove. How cleanly channel markers
lean to their bell-rocked time and the foghorn urging:

Not to worry, safe—safe. The waters just a moment chill,
        the current lulled—a mother's rocking. Three or four
strokes from shore, just far enough to see how fog shoulders
        the headland in white. Then the eddying pull of a red buoy,
the sweet rip of current furling, easing—*there now, there.*

Broodmares Staring

Last year's bleached corn breaking through
        light snow—pleasing rill, grace for the eye.

Imagine a cloth thrown over a feast—salt shakers
        become a procession of kings; forks and spoons

and stacked plates—ridged Babylonian gardens
        bordering water. Just so—remnants

of cornstalks run the lines of hillside,
        lush green of summer ghosting now

in diminished height, translated, resurrected in white.
        We walk on through the chill hollow

to the top of the rise, just shy of broodmares
        lazily looking up: a wind stir of blown snow—

and no tracks. We might be anywhere; these winter walks
        a slow waltz through the first house before children—

each new room, rhythm of the field—hill, gully,
        rib, hip, each bright particular sloping away.

Night Feeding

One-thirty, and the owls call cedar to pine
threading in and out of hedges, over lilac,
under larch. And you, small wren, root and suck
in the hawthorn of my nightshirt. Time contracts, the axis
poled at this hour, in this room, constellations unfurling.

We are unfallen, mother and child in a prelapsarian wild
like the Renaissance painting I dreamt the morning
you were born: verdant landscape, folded and lush
as unnapped velvet, a rough path winding, hills
onto hills, streams running but all of it still.

Hushed as compared with the Madonna's eyes,
the child's gaze. In that look flesh quickened.
The one square frame suddenly breathing.
And in the dream, I turned looking for someone
to witness the brightness—the mother's singing,

the child's eyes following hers, moon lucent, canny.
But there was no one there and the heaven-lit,
breathing square receded into brushstroke and dark oil.
The warmth of lips and cheeks gone pale
as winter yew berries. The vision just a moment real.

And that's how it is at this feeding but no other.
For by five a.m. the suburb has regained its boundaries.
What might have been a meadow is a cropped
yard. The paper boy thwacking each missive
in a separate drive. First light whitens—

the owls have lost one another—
it's only the mourning doves, melodic yet naive.
No one will know we stood among these trees,
in darkness rich and strange, coveting only
one another's gaze, your breath sifting with mine.

Still-Life Breathing

Vast—it seems the figures move.
Rake of light, a hem lifts
and Cavallini's angels stir.
Like hay in wind their hues shift.

Base of the feathers, a faint
pink, rinse water of beets
as you loosen their skins.
Incrementally, color increases

with each row of feathers,
slowly rose goes sage-green.
It's enough to make you believe.
Leaves unfurling mid-winter,

the sun's arc steep.
This is where they brought her.
Not at first, but later;
for Santa Cecilia's body was lost

in the catacombs of San Callisto,
the carved out underearth of pumice, ash
and lava where she lay as if sleeping:
knees bent, hands unclasped.

"An angel," she told her husband
on their wedding night,
"an angel guards my purity."
He believed in what he couldn't see.

Imagine forfeiting the rushed
fire of husband and wife:
lip, brow, eyelash—a fern's lithe
shiver, inert upon your cheek.

Tableau: After the Storm

I.

Not a scrap of sail idling on boom or mast
    To trouble or remind the eye. Just sand and sea
Where once the beach house stood. All the past
    Windswept, destruction so seamless

It might have been dreamt—like the boyish cries
    Of selkies in each reaching wave. Along Coatue
Terns nest in brittle creels of weed and debris, tides
    Rise and fall away. A dark watermark eases through

The jetty's black rock—but all of it's skewed.
    Here at the water's edge the sea wall stood,
Roses fell spilling on forget-me-not and bluet.
    Up on the bluff the Bredin girls slept,

Windows open—curtains wild as spinnakers loosed.
    Nothing left to tell of three sisters stealing, knee
Over wrist, light-headed, reckless in the bruised
    Surf before dawn. Latches cast, ponies freed.

Not a mark where lovers made angel wings in sand
    Or a striped tent billowed and rolled phantom
In gusting rain. There—at the bluff—the youngest wed
    Clear browed as smoothed sand under the thrum

Of an August-end squall. And the eldest wandered,
    Out from the wedding, into the wind. Not lost
In envy as the others surmised, despite herself grieved
    In the presence of bliss, but drawn by the lines

Of seals, bright-eyed familiars come again inland.
    Lithe and compelling as the noise of lost children.
And so beguiled she ran in her silk dress, diving to meet them.
    But now, in the storm's wake—just thin air and light

Where the garage stood and they all danced, where a summer
    Later the sparrow hawk lived, skittish as moonlight
When the backdoor slammed. And caught: a stammer
    Of feather and claw, bronzed in fright.

2.

Dusk and tide and ribbed sand. Channel-keepers bell
    The quiet, heeling into current, oblivious
Of the changed shore. The past no more than name might tell,
    Duneover, stripped dunes tumbling seaward, raucous

Terns. Even so quick, memory becomes rumor,
    Something fishermen knew and children murmur:
The wild horses in Tuckernuck's surf, the blue-gray
    Seal skins drying off Muskeget, the fey

Unearthly singing—just the wind in the ear of a whelk.

Salt

I am not the one you think, I am
        A kept thing loosed, the sleeping vellum underside
Of an unsealed thought—that which you forgot,
        The wren who flew in at the open door.

I am not her—your wife, the children's mother.
        Watch me now closely—where she moved
Clockwise round the table, I'm at it withershins,
        Spooning out the wild rice and snow peas.

        Even the neighbors are fooled as I stretch
And root in cool earth, making do
        With white grubs curled like sleepers.
The oniony bulb at the grass tip's root.
        They assume I am her leaving off verse
For the garden again.

I am not the one you kissed like the last china cup.
        I had no part in the laying out of breakfast,
        The oysterwhite bowls, something for each:
Comics for the willowy one, wolves for the red-haired girl,
        A pirate's eyeglass for the boy.

        I am the spoon lapped with milk,
        A thin thread of snowmelt negotiating shale,
I came in on your words—
        You remember this—

Laying hands on the shells lining the sill,
Choosing first the scallop, then laying it down
        For the bleached white clam,
The sweep of your hand assessing the fine nub—

    And here you paused,
    What you meant to say was that there was
No room on the sill for your keys what with all those shells...
    But in that breath—air met ocean,
    The great gate unhinged
    And I let myself in.

I am not her, but I am yours,
    Like salt for meat I thirst for water.
Deprived of ocean—I seek out rainfall, taps.
    I fill the dog's water bowl over and again,
Let the water run. The spilling returns me.

I came in on your words
    And will leave on hers.
Until then I'll lie beside you and in the morning
    Wake the children one by one,
Lead them downstairs, my shoulders round their own
    Like a light vest, like angels' wings
Their lovely elbows cupped in my palms.

Dwelling

The outside is in you now—
that Japanese maple too bright
to be believed in on this dark day.
Does the rain make you wait?
The skittery path of a morning
moth? It flies too low and slowly.
More likely company, white pine and cedar
keeping the saw-whets from grief.
You'd never know they're there
but for their mewling calls. Just a day,
a night, a day as they make their way
south, southwest to the wintering places.

I never told you about the summer moth
wide as my palms, auburn shading
to doe-brown, fringe of pink
fading to a blue luna line.
I found her under the new pear,
wings flush with the summer grass.
Heartbroken or dying, so it seemed
until she shuddered eyespot to eye,
a lidless waking on the powdery
slope of wing. Two small rondels
see-through as cellophane, clear
as prayers, our world to yours.
Lying flat, warmed by the earth,
there's just a glimpse of grass.
But imagine them in flight—
Antheraea polyphemus—the moving blue

eye of sky caught in the round pane,
all the places you might be hidden;
I'll wait for the color, watch for the wing.

Into the Blue

Lure of a glass-quick chromis, like something stolen
from the night sky. Striped eye of a spotfin

passing once, drawing back, a dorsal fin's deft
maneuvering. Something in me wakes, and I follow—

ochre, umber, alizarin crimson—
until I've lost them. There's only the slow wrap

of sea fan, antlers of elkhorn—this sudden blue chill,
like unearthly music in the mountains, goat-footed

unaccountable clamor, and I see where I am,
a hundred shimmering feet deep, the blue below

a solid and impassive vault. I breathe water not air
in a black-flippered spin; I am foolish and naïve and far

from the others—until I see them, a fleet of spotted blue damsels,
all the light-aqua eyes on their indigo sides revolving.

Then a silver school of bar jacks pass through me,
And I fire the lime-white mounds like loosed eels—blush-pink,

star-studded, branching and scrolled. The dark out there
is my own; a scarlet anemone is weeping outward for me.

And I know where I am. Time's so slow here
you could scoop up any of them—triggerfish, squid,

rock hinds, and hamlets—even the chaste blue chromis.
Another salt breath and I'd fold him to the shadowy pale

between my breasts. Skin to scale, eye to eye,
in a shifting bright dance. But there'd be hell to pay

for the sleek wild of a trespassing instant.
Even if I flipper-kick like light itself for air,

I'll never rise knowing what they know, never keep
the unthinking grace. Just breathing air is breaking

all kinship. Ascending, thinking, breathing
and it's gone. There's only the quick falling dark,

a high wind rising. And at my fin's tips
a comet's tail.

Winter Abundance

A half-apart mower
and a seam-cracked feed bucket.
Two bare wooden chairs
facing the wood. Bucket of screws,
and a bag of old golf clubs;
there's a chancy grace here.
The deer don't mind
and the horses take comfort
in disrepair so generous
that no one thing
could scut an alarm
tripping through dry leaves.
Even the crossbow, north by
northwest, points like an angel
inclined for the pines.

Last year's white antlered buck
is still ghosting his way,
Seahorse to Mid-Stream.
He's one of us in the dark
of February, keeping at it.
He straddles the berm
to watch the thoroughbred
buck, fuss, and revel—
in the joyous cold
of the thin blue hours.
Just as sudden, the back
of havoc sundered—
wild canters grown smooth,
smoother. Long hours

grown simple with work:
dozens of eights & loops,
habitual refining serpentines
in the hold of the wood—
a settled low contentment—
at one with the farm shed
strewing its splendor like snow.

Seafarer

That's Walt Anderson napping in the narrow berth
of the skiff, arms folded neat as a bishop.
The heat of the day settles into the ribbed place.
Lift the seat boards, slip the oars—
it's like being a cloud in the wind's swift carry,
tossed around, blown about and all the while
sun beats warmth on grizzled beard and cheek.
I don't remember when it started, this habit of his,
afternoon naps in the belly of a boat—
when we were young, I suppose.
Long summer nights out on the bay,
we'd feel our way along the lines, hand
over hand we hauled in, seawater spilt like light.
Lobster and crayfish, now and then the odd catch:
sailor's button in a flush of seaweed, a wayward perch.
Some nights, we'd light fires on the beach—
driftwood blaze and late supper, clams on hot rocks,
seaweed snapping in heat. Once we brought girls,
someone's visiting cousins not well accounted for;
wet bed of seaweed, salt mist. He's dead asleep he is,
and it won't be long before the old Norseman heads on.
I'd slip the anchor this minute if I thought he were close,
send him off dreaming, hauling in lobster.
The sound of surf breaking over that hiss-and-spit
ruckus we've always called the hay ledge.
We rode it standing up, hundreds of times—
hallooing and whooping as waves hit and slipped,
and once in a while our skiffs would lift
wave on wave over the long white air.

Watchers

Even the windows in their frames—the roughshod
ride of new paint over the old wood, blisters
spilling in thick bands, then shallowing like pools in rock—
would do for him, let alone the pond, smallest and darkest

of the three adjoining waters—there, just outside.
It was lonely enough and calm, familiar somehow.
He could see things here, legs crossed on the stool
by the door; the blood-ruby glint of double-decked

wings, bead-by-bead composure of a dragonfly's
slim thorax and at the border, the narrow head
and piercing, dark-leafed stare. Curious, yet disinterested
student of the quiet. His own twin at the pond's edge

memorizing each shivery twist of sun though water-shield,
the shifting bars of perch and then, the too still greenness
of a frog's sides. The water snake's arc through water,
the hapless, headfirst, into the sprung jaw plunge—swift

as his own thought. Race and strike, followed by the long
unseemly swallowing: belabored, even comic
as the slithering body, headed by the backwards leaping
hind end of a frog makes slow to no progress in the cattails.

Like the shape of the thing taking shape at his knee:
Beach grass and indigo weed, fox in a hollow
with a white-tipped tail, fog making pasture
of a turned-up sea—*he liked the beach peacooked green*—

the unordered rush of it fleshing out, coming clean
in the quiet of the empty house by the open door.

After Thoreau, by Williams pond

All Souls Crossing

Black pond coming off the sturdy pines
Through shaking banks of phlox
To the simple white cottage. Why not here?
All of us queuing up politely by the screen door
Until we remember, as at first it's hard to remember,
The habit of souls no longer clothed in bodies so new.
Will it be broad as a whale path, the crossing?
Or thin as willow bark strips, dark stature
Of your father, fire in hand, rising to my grandfather's knock,
Your brother a boy again under a thatch of bangs
Barefoot on a flying chase. Think of it now
Before white clouds drum and you haven't a minute
As soul flies its quick heel through slender squares
Of window screen, into loose whispers of hide & seek
Under beach plum, lawn chairs, crazy-eighting
In and out of house and garden, the shoved scud
Of an upstairs bed sliding, a slammed door's catch and hit.
Quick, slip into the garden's musk-rut, deliberate
As a dozen honey bees at work, headfirst into blue throats
Of catnip. Stock-still under capes of sun-flowers,
A white-throated sparrow balances, his one song
Repeating in the brassy ever-changing light
Before earth goes and the white curtains pressing
To window and jamb are suddenly still.

Where the Hill Rises

There was another life before yours
in these old stone rooms,
two opposing fireplaces keep watch
like the staring greyhounds flanking
the door. Privet, vinca, long spears of lavender.
One wooden chair faces the sun,
another angles off under the old pear,
twinned and unlikely as pearl earrings
and a black purse of skate eggs
trapped between window & screen,
forgotten all winter.

Someone else slept in these rooms,
left the bed without your knowing,
wandered down to the stable
to breathe in the dark swill,
piss and straw and deep contentment.

With Trawlers and Fishermen

A small boat stutter-shims in light surf, rocks a bit
but goes nowhere, so answerable to gesture
and weight that one foot in the bow
can send it heaving—that is the way
of our summer bed in the small white room.
Wooden pegs for nightshirts, tee-shirts, shorts,
fine blue curtains slim as loose jibs, just enough
hip room for a lamp and book. This bed's
too small for separate sleep, one ghost glance—
we are tinder again, thin-flanged greens and snapping
twigs in smoking sleeves, sharp with the damp,
then fiery, limbstruck—rock oak, woodbine,
witch hazel, heath, sassafras, sweet vetch,
bayberry, spruce. From here to the Atlantic
our watchfires keep—'til wind raps the glass
in the old sills and a mournful whinny
calls us home from the scattershot brights
to low roof and orchard, red-right-returning
with trawlers and fishermen we make our way in,
salt-resined and creased, loose limbed
as morning wind beating a path through white pine.

Notes

The first epigraph is from Wendell Berry's "The Old Man Climbs a Tree," in *A Timbered Choir: The Sabbath Poems;* the second is from *Two Worlds of Andrew Wyeth: A Conversation with Andrew Wyeth* by Thomas Hoving.

"Flying Change": the airborne switch of leads happens at a canter or gallop. When done well, the transition is nearly seamless. I first encountered the term used as a title in Henry Taylor's third volume of poems.

"Into the Green": variants on the pastoral; the title is derived from, but not specific to, what Northrop Frye called "the green world."

"Paint Like Breath on a Pane of Glass": adaptation of James McNeill Whistler: "Paint should not be applied thick. It should be like breath on the surface of a pane of glass."

"Fear of Heights" and "Seafarer" take inspiration from Andrew Wyeth's *Widow's Walk* and *Adrift* respectively. "Anna Kuerner" draws on *Groundhog Day;* Wyeth's extensive studies for this painting were useful in thinking about themes of disappearance and transfiguration.

"At Tea, Far Side of the Glass": the italicized lines are from Edna St. Vincent Millay's "Childhood Is the Kingdom Where Nobody Dies."

"Edgeless and Smoothed Out": the title is from Charles Wright's "Buffalo Yoga."

"Thinking About the Sistine": the italicized phrase is from a poem Michelangelo was writing while painting *The Last Judgment*. The translation is by Frederick Hartt.

"The Breathing Green: an Elegy": is indebted to Richard Meryman's *Andrew Wyeth: A Secret Life* for details about N.C.'s accident. The italicized lines are from Meryman.

"Mother, Daughter...": the broken statue of the Madonna was unearthed in a nearby vegetable garden. Locals refer to Santa Maria dei Miracoli as "the jewelbox."

"Still Life Breathing": Pietro Cavallini's angels appear in the upstairs frescoes of the church of Santa Cecilia. St. Cecilia was martyred in the house beneath the church.

"Watchers": Thoreau spent the night at a Wellfleet oysterman's cottage along the eastside of Williams Pond. The italicized phrase is from Thoreau's *Cape Cod*.

"Red-Right-Returning": a mnemonic expression used to remember the navigational rule that, when returning inland, you must keep red buoys on the right.

"Winter Abundance": the italicized lines are from Shakespeare's *As You Like It;* for their significance to the pastoral, I am indebted to Wendell Berry's the "Uses of Adversity."

McGovern Prize Winners

Catherine Staples, for *The Rattling Window* (nominated by Eamon Grennan)

Robert Grunst, for *Blue Orange* (nominated by Marilyn Chin)

Christine Gelineau, for *Appetite for the Divine* (Editor's Choice, selected by Deborah Fleming)

Elizabeth Biller Chapman, for *Light Thickens* (nominated by Enid Shomer)

Michael Miller, for *The Joyful Dark* (Editor's Choice, selected by Stephen Haven)

Maria Terrone, for *A Secret Room in Fall* (nominated by Gerry LaFemina)

Nathalie Anderson, for *Crawlers* (nominated by Eamon Grennan)

A.V. Christie, for *The Housing* (nominated by Eamon Grennan)

Jerry Harp, for *Gatherings* (nominated by John Kinsella)